Catoptric Conversations

Reflections on Light

Waldo Hersey

Chapter 1: The Language of Light

Photons and Perception

We perceive the world as a vibrant tapestry of colors, shapes, and textures. Yet, this reality, this sensory experience, is nothing more than our brain interpreting the language of light. Imagine a world shrouded in perpetual darkness. It's a chilling thought, isn't it? Without light, our very understanding of existence would fundamentally shift. We wouldn't experience the world as we know it. This is because light, in its purest form—photons— serves as the messenger of our reality.

These massless particles, traveling at an astonishing speed, carry information from the far reaches of the universe to our very eyes. When we talk about seeing, we are actually talking about photons interacting with the world around us and ultimately, with our senses. Think of a single photon embarking on a journey, perhaps bouncing off the petal of a rose. It carries with it a tiny packet of energy, a specific wavelength that our brains interpret as the color red. This seemingly simple act, repeated billions of times over, allows us to perceive the beauty and complexity of our surroundings.

But our perception of light goes far beyond simply identifying colors. The way light interacts with objects—how it reflects, refracts, and diffracts— informs our understanding of shape, depth, and texture. The subtle play of light and shadow on a sculpted face, for instance, allows us to perceive its contours, its emotions. It's this interplay between light and form that artists have captured for centuries, using techniques like chiaroscuro to create an illusion of three-dimensionality on a flat canvas.

And consider the way sunlight filters through a forest canopy, dappling the forest floor in a mosaic of light and shade. This interplay isn't just aesthetically pleasing; it plays a crucial role in the life cycles of plants and animals. Plants have evolved to thrive in specific light conditions, their leaves acting as solar panels, capturing the energy of photons to fuel photosynthesis. Animals, too, rely on light for navigation, communication, and even camouflage. The bioluminescent glow of a firefly on a summer night is a testament to the power of light to shape life itself.

Our perception of light is also deeply intertwined with our emotions and psychology. Bright light can evoke feelings of joy and alertness, while dim lighting might inspire tranquility or even melancholy. Think of the warm glow of a fireplace on a winter night, the comforting feeling it evokes. Or the stark, sterile light of a hospital room, often associated with anxiety and unease. These emotional responses to light are not merely cultural constructs; they are rooted in our biology. Our circadian rhythms, the internal clocks that regulate our sleep-wake cycles, are heavily influenced by the presence or absence of light.

But perhaps the most profound aspect of light is its role in shaping our understanding of the universe. For centuries, astronomers have relied on light, traveling across vast distances, to study celestial objects. By analyzing the spectrum of light emitted by stars and galaxies, we can determine their composition, temperature, and even their movement. The faint afterglow of the Big Bang, known as the cosmic microwave background radiation, provides us with a glimpse into the earliest moments of the universe's existence.

Light, in its many forms, is fundamental to our existence. It shapes our perception of the world, influences our emotions, and guides our scientific exploration. As we delve deeper into the mysteries of light, we uncover not only the secrets of the universe but also a deeper understanding of ourselves.

A Spectrum of Stories

Light, in its vast and varied spectrum, holds within it a universe of stories. From the soft glow of a firefly illuminating a summer night to the brilliant explosion of a supernova marking the end of a star's life, light narrates the unfolding drama of the cosmos. Each photon, a tiny messenger carrying information across time and space, whispers tales of distant galaxies, fiery births, and quiet deaths.

Consider the humble act of looking up at a starlit sky. Each pinprick of light, a distant sun, perhaps harboring its own system of planets and possibilities. Some of these stars, we now know, have planets orbiting them, worlds bathed in the light of their own suns. What stories might unfold on those distant shores? What forms of life might bask in the glow of alien stars? The light that reaches us, having traveled for eons, offers a tantalizing glimpse into these cosmic narratives, sparking our imagination and fueling our sense of wonder.

But the stories held within light are not limited to the grand scale of the universe. They exist all around us, woven into the fabric of our everyday lives. Think of the way sunlight, streaming through a window, can transform a mundane interior into a scene of quiet beauty. The way dust motes dance in the golden rays, each one a tiny microcosm of swirling energy. Or the way light, filtering through stained glass, paints a cathedral floor in a kaleidoscope of colors, imbuing the space with a sense of awe and reverence.

These stories are not always told through visual spectacle. Sometimes, the most powerful narratives are whispered in the subtlest of ways. The soft glow of a bedside lamp, for instance, can evoke feelings of comfort and security, reminding us of childhood nights spent reading under its warm embrace. The flickering light of a candle flame can create an atmosphere of intimacy and reflection, inviting us to slow down, to connect with the present moment.

Light also has the power to reveal hidden stories, to illuminate aspects of the world that might otherwise remain unseen. Think of the way a photographer, using light and shadow, can capture the essence of a person's character in a single, fleeting moment. Or the way a scientist, peering through a microscope, can unlock the secrets of the natural world, revealing the intricate structures and processes that govern life itself.

The stories told by light are not always straightforward; they can be fragmented, ambiguous, open to interpretation. Think of the way a painter might use light and color to evoke a particular mood or emotion, leaving the viewer to piece together the narrative for themselves. Or the way a filmmaker might use lighting techniques to create suspense, intrigue, or a sense of the uncanny.

In a sense, we are all storytellers of light. Every photograph we take, every image we share, every time we illuminate a space, we are shaping the way light interacts with the world, creating our own unique narratives. These stories, in turn, have the power to connect us, to inspire us, to challenge the way we see ourselves and the world around us.

So, the next time you find yourself bathed in the warm glow of the setting sun, or captivated by the twinkling expanse of a starlit sky, take a moment to consider the stories that light has to tell. For within its radiant embrace lies a universe of wonder, waiting to be explored.

Illuminating the Invisible

Light, in its dance of reflection and refraction, has the remarkable ability to illuminate not just the visible, but also the invisible. It allows us to peer into the hidden depths of the microscopic world, revealing the intricate structures and processes that underpin life itself. It also grants us access to the vast expanse of the cosmos, unveiling celestial objects and phenomena that lie far beyond the reach of our unaided vision.

Consider the microscope, an instrument that has revolutionized our understanding of biology and medicine. By harnessing the power of light, we can magnify objects thousands of times their actual size, revealing a universe teeming with activity that exists beyond the limits of our perception. Cells, once thought to be simple building blocks of life, come alive under the microscope's lens, their intricate machinery of organelles and molecules engaged in a constant dance of life and death.

But the microscope is not merely a tool for observation; it is also a powerful instrument of discovery. The use of fluorescent dyes, which emit light when excited by specific wavelengths, has allowed scientists to track the movement of molecules within cells, to visualize the complex interactions between proteins and DNA, and to gain a deeper understanding of the mechanisms of disease.

Light's ability to illuminate the invisible extends far beyond the microscopic world. Telescopes, from the modest backyard variety to the massive observatories perched atop mountains, capture light from distant stars and galaxies, allowing us to peer billions of years into the past. Through the lens of a telescope, faint smudges of light resolve themselves into swirling galaxies, each containing billions of stars. We can witness the birth of stars in glowing nebulae, the death throes of massive stars in supernova explosions, and the enigmatic dance of matter around black holes.

Moreover, by analyzing the light emitted by these celestial objects, astronomers can decipher their composition, temperature, and movement. Spectroscopy, the technique of splitting light into its constituent wavelengths, reveals the unique fingerprints of elements present in stars and galaxies. These spectral lines, like cosmic barcodes, provide clues about the processes occurring within these distant objects, helping us to piece together the history and evolution of the universe.

But light's ability to reveal the unseen is not limited to scientific instruments. Artists, too, have long understood the power of light to illuminate the invisible, to make the intangible tangible. Think of the way a painter might use light and shadow to convey the depth of a landscape, the texture of a fabric, or the emotions flickering across a subject's face. Or the way a photographer, by manipulating light and exposure, can capture the fleeting beauty of a moment, revealing the hidden patterns and rhythms of the world around us.

Even in our everyday lives, we experience the power of light to illuminate the invisible. Think of the way sunlight, streaming through a window, can reveal the dust motes dancing in the air, particles we might not otherwise notice. Or the way a beam of light, shone across a darkened room, can reveal the intricate details of a spiderweb, its delicate threads shimmering like spun gold.

Light, in its many forms, has the remarkable ability to make the invisible visible, to bridge the gap between the seen and the unseen. It allows us to explore the hidden depths of the microscopic world, to peer into the vast expanse of the cosmos, and to appreciate the subtle beauty that surrounds us in our everyday lives. By embracing the power of light, we open ourselves up to a deeper understanding of the universe and our place within it.

Chapter 2: Mirrors of the Mind

Reflections of Self

We gaze into the mirror, and what stares back? Is it merely our physical form, a collection of features and expressions? Or does the reflected image reveal something deeper, a glimpse into the essence of who we are? Throughout history, mirrors and reflections have held a certain mystique, serving as potent symbols in mythology, literature, and art. They invite us to contemplate our sense of self, our identity, and our place in the world.

On a purely physical level, a mirror reflects light back to our eyes, creating an image of ourselves. Yet, this seemingly simple act has profound implications. The act of seeing ourselves, of recognizing our own reflection, is a crucial step in developing a sense of self-awareness. As infants, we first learn to distinguish ourselves from the world around us by recognizing our own image in a mirror. This recognition, often accompanied by smiles and excited babbling, marks a significant milestone in our cognitive development.

But the significance of reflections extends far beyond the realm of child development. Mirrors, both literally and metaphorically, provide us with opportunities for self-reflection and introspection. We gaze into the mirror not just to check our appearance, but to examine our thoughts, feelings, and motivations. The face that stares back at us can reveal a range of emotions—joy, sorrow, anger, fear—reflecting our inner state of being.

Artists throughout history have explored the power of reflections to convey complex emotions and ideas. Think of the haunting self-portraits of Rembrandt, in which the artist's weathered face seems to reflect a lifetime of experiences, both joyous and sorrowful. Or the surrealist works of Frida Kahlo, in which mirrors are used to explore themes of identity, duality, and the fragmented nature of the self.

Literature, too, is replete with examples of reflections serving as metaphors for self-discovery and transformation. In Narcissus's fateful encounter with his own reflection in a pool of water, we see the dangers of vanity and self-obsession. In contrast, Alice's journey through the looking glass offers a fantastical exploration of identity and the fluidity of perception.

But reflections are not always about solitary introspection. They can also serve as powerful reminders of our interconnectedness with others. When we see our reflection in the eyes of another person, we catch a glimpse of ourselves through their perspective. This shared gaze, this moment of mutual recognition, can foster empathy, understanding, and a sense of shared humanity.

The reflections we encounter in the natural world also offer opportunities for contemplation and connection. The still surface of a lake, mirroring the sky above, reminds us of the interconnectedness of all things. The way sunlight, reflecting off a spider's web, transforms it into a shimmering tapestry of light, invites us to appreciate the beauty and wonder that exist in the everyday world around us.

Ultimately, reflections, in their various forms, challenge us to look beyond the surface, to delve deeper into the complexities of self and identity. They remind us that who we are is not fixed or static, but rather a constantly evolving reflection of our experiences, relationships, and interactions with the world around us. So, the next time you find yourself gazing into a mirror, take a moment to appreciate the multifaceted nature of your own reflection. For within its depths lies a universe of stories, waiting to be explored.

Distortions and Delusions

The world, bathed in light and shadow, can be a deceptive place. Our perception, honed by evolution to navigate a world of physical objects and tangible threats, can be easily fooled, tricked by illusions that exploit the very shortcuts our brains use to make sense of the visual world. Distortions and delusions, those quirks of perception where what we see diverges from objective reality, offer a fascinating glimpse into the intricate workings of the human mind.

Consider the humble mirage, a shimmering oasis appearing in the distance of a scorching desert. This illusion, born from the bending of light rays as they pass through layers of air with varying temperatures, has lured countless travelers off course, their thirst-addled minds mistaking the shimmering reflection for a life-saving source of water. The mirage, while ultimately an illusion, reveals the brain's powerful drive to find patterns and meaning, even in the most chaotic of environments.

But distortions are not limited to the natural world. Optical illusions, those carefully crafted images that play tricks on our visual system, demonstrate the complex interplay of light, perception, and cognition. The impossible objects of M.C. Escher, with their gravity-defying staircases and endlessly looping pathways, challenge our assumptions about perspective and three-dimensional space. These illusions, while playful in nature, highlight the fact that our perception of reality is not a passive process, but an active construction, a collaboration between the information our eyes receive and the interpretations our brains create.

Delusions, unlike illusions, represent a more profound disconnect from reality. These firmly held beliefs, often unshakeable even in the face of contradictory evidence, can manifest in a variety of ways, from the relatively benign to the deeply disturbing. The gambler who believes he's on a winning streak, the conspiracy theorist who sees patterns in random events, the individual who believes they are being persecuted by unseen forces—all demonstrate the power of belief to shape our perception of the world.

While the causes of delusions are complex and multifaceted, they often involve a disruption in the brain's ability to process information effectively. Trauma, mental illness, substance abuse, and even extreme stress can all contribute to the development of delusional thinking. Understanding these underlying causes is crucial for developing effective treatments and interventions.

Yet, even as we strive to understand and address the more troubling aspects of distorted perception, we can also appreciate the beauty and wonder that these quirks of the mind reveal. The shimmering colors of a soap bubble, the mesmerizing patterns of a kaleidoscope, the awe-inspiring illusions created by stage magicians—all remind us of the power of perception to transform the mundane into the magical.

Ultimately, our understanding of the world is shaped by the interplay of light, perception, and cognition. While distortions and delusions highlight the potential for error and misinterpretation, they also reveal the remarkable adaptability and creativity of the human mind. By embracing the full spectrum of human perception, from the objective to the subjective, we can gain a deeper appreciation for the complexities of consciousness and the multifaceted nature of reality itself.

The Search for Authenticity

In a world saturated with images, where reality itself often seems filtered and curated, the search for authenticity has become a defining pursuit for many. We yearn for genuine experiences, for connections that run deeper than surface-level interactions, for a sense of self that feels true and uncontrived. But what does it really mean to be authentic in a world of carefully constructed personas and ever-shifting trends?

The concept of authenticity is complex and multifaceted, often intertwined with notions of self-awareness, integrity, and vulnerability. It's about being true to our values, even when they differ from the norm. It's about embracing our imperfections, recognizing that our flaws are part of what makes us unique. And it's about connecting with others on a deeper level, sharing our authentic selves without pretense or artifice.

The pursuit of authenticity, however, can be fraught with challenges. We live in a culture that often rewards conformity and self-promotion, where the pressure to present a polished and idealized version of ourselves can feel overwhelming. Social media, with its carefully curated feeds and highlight reels, can exacerbate this pressure, leading to feelings of inadequacy and a sense that our own lives pale in comparison.

Yet, it's within this very tension—between the pressure to conform and the desire to be true to ourselves—that the search for authenticity takes on its greatest significance. It's about finding the courage to resist the urge to conform, to embrace our individuality, and to let our true selves shine through.

This process of self-discovery often begins with introspection, with taking the time to reflect on our values, beliefs, and motivations. What truly matters to us? What makes us feel alive and fulfilled? These are not always easy questions to answer, but they are essential ones to ask if we want to live authentically.

Authenticity is not about achieving some fixed state of being; it's an ongoing process, a journey of self-discovery and growth. It's about being present in each moment, engaging with the world around us with openness and curiosity, and allowing ourselves to be changed by our experiences.

One of the most powerful ways to cultivate authenticity is through genuine connection with others. When we feel safe and accepted for who we are, when we can let down our guard and be vulnerable, we create the space for authentic relationships to flourish. These connections, built on trust, empathy, and mutual respect, provide us with the support and encouragement we need to navigate the complexities of life and to continue evolving into our most authentic selves.

The search for authenticity is not a solitary pursuit. It's a journey that we undertake within a community, drawing strength and inspiration from those around us. By embracing our individuality, by connecting with others on a deeper level, and by living in alignment with our values, we can create a more authentic and fulfilling life, both for ourselves and for those around us.

Chapter 3: Through a Glass Clearly

Lenses of Perception

We don't see the world as it is, but as we are. This profound statement, often attributed to the philosopher Immanuel Kant, speaks to the powerful influence our individual perspectives have on shaping our reality. Each of us experiences the world through a unique set of lenses, molded by our upbringing, experiences, values, and beliefs. These lenses, while essential for navigating the complexities of life, can also color our perceptions, influencing how we interpret events, interact with others, and ultimately, experience the world around us.

Imagine for a moment, a bustling city street, teeming with life. A street musician fills the air with melodies, their music interpreted as joyful by some, melancholic by others. A child's laughter dances on the breeze, a sound that might evoke feelings of warmth in one passerby, a twinge of longing in another. The same scene, witnessed by multiple individuals, can evoke a kaleidoscope of emotions, interpretations, and reactions, each filtered through the unique lens of personal experience.

Our upbringing plays a pivotal role in shaping these lenses. The family we are born into, the culture we are raised in, the values and beliefs instilled in us during our formative years—all leave an indelible mark on our worldview. A child raised in a loving and supportive environment might view the world with a sense of optimism and trust, while a child who has experienced trauma or neglect might approach life with caution and suspicion.

As we navigate the world beyond our childhood homes, our experiences continue to shape and refine our lenses. Every interaction, every challenge overcome, every heartbreak endured adds another layer of complexity to our perception. A successful entrepreneur, having weathered countless risks and setbacks, might view failure as an opportunity for growth, while someone who has experienced repeated setbacks might approach new ventures with trepidation.

Our values and beliefs act as filters, influencing what we pay attention to, how we interpret information, and ultimately, what we believe to be true. Two individuals with differing political ideologies, for example, might interpret the same news story in vastly different ways, their pre-existing beliefs shaping their understanding of the events.

These lenses of perception, while integral to our individual identities, can also create barriers to understanding and empathy. When we view the world solely through our own limited perspective, it becomes easy to judge, misunderstand, and misinterpret the actions and motivations of others.

However, by becoming aware of our own perceptual lenses, and by acknowledging the validity of other perspectives, we can begin to bridge these divides. Cultivating empathy, that ability to step outside of our own experience and see the world through another's eyes, becomes paramount.

Engaging in open and honest dialogue, listening with the intent to understand rather than to respond, and seeking out diverse perspectives can help us broaden our own lenses, allowing for a more nuanced and compassionate understanding of the world and its inhabitants.

The beauty of the human experience lies in its diversity. By recognizing and appreciating the myriad lenses through which we each experience the world, we open ourselves up to a richer, more meaningful existence. For it is in the spaces between our individual perceptions, in the shared exploration of our common humanity, that true understanding and connection can flourish.

The Illusion of Objectivity

Objectivity, that elusive ideal of unbiased perception, holds a revered place in our understanding of truth and knowledge. We strive for it in journalism, science, and even in our personal lives, believing that by stripping away our subjective biases, we can access a purer, more accurate view of reality. But what if this pursuit of objectivity, particularly when observing the human experience, is itself an illusion?

The human mind, far from being a neutral observer, is an active interpreter, constantly shaping and making sense of the world through the lens of individual experience, cultural conditioning, and a myriad of unconscious biases. We don't absorb information passively; we filter it, categorize it, and compare it to our existing frameworks of understanding. This inherent subjectivity colors every aspect of our perception, making true objectivity an aspirational goal rather than an attainable reality.

Consider the very act of observation. In quantum physics, the observer effect demonstrates that the mere act of observing a phenomenon can alter its state. While this concept applies to the subatomic realm, it holds a powerful metaphor for understanding human perception. Our presence, our attention, our intentions—all influence the very thing we seek to observe objectively.

Furthermore, language, the tool we use to articulate our observations, is inherently subjective. Every word carries a nuanced history of cultural and personal associations, shaping not only how we express our thoughts but also how we think about the world around us. Even seemingly neutral descriptions can carry subtle biases, influencing our understanding and interpretation of events.

The illusion of objectivity is further compounded by the human tendency to seek patterns and narratives. Our brains are wired to find meaning and coherence in the world, often leading us to perceive connections and causality where none may exist. This innate desire for order can lead to confirmation bias, where we selectively seek out information that confirms our pre-existing beliefs while dismissing contradictory evidence.

Even scientific inquiry, often held up as the gold standard of objectivity, is not immune to the influence of subjective biases. The questions we choose to ask, the methodologies we employ, the interpretations we draw from data—all are influenced by our theoretical frameworks, our cultural contexts, and even our personal values.

Recognizing the illusion of objectivity is not about abandoning the pursuit of truth or embracing relativism. Instead, it's about acknowledging the inherent limitations of our perception and approaching knowledge with a healthy dose of humility and critical self-reflection.

By understanding the various filters through which we experience the world, we can begin to deconstruct our own biases, challenge our assumptions, and cultivate a more nuanced and compassionate understanding of ourselves and others. True insight, then, lies not in denying our subjectivity but in embracing it, in recognizing that our individual perspectives, while inherently limited, form the rich tapestry of human understanding.

Chapter 4: Prismatic Perspectives

Refracting Reality

Reality, once considered a fixed and immutable concept, now appears increasingly fluid, refracted through the prism of our individual perceptions, shaped by the relentless torrent of information, and molded by the very technologies we create. We inhabit a world where the lines between the physical and the digital, the real and the virtual, blur with increasing frequency, challenging our assumptions about the nature of truth, experience, and even our own identities.

The rise of social media, with its carefully curated feeds and echo chambers, has profoundly impacted how we perceive and interact with the world. We are bombarded with a constant stream of information, much of it filtered through algorithms designed to reinforce our existing biases. This creates a self-perpetuating cycle, where our perceptions of reality become increasingly narrow and polarized, shaped more by the digital echo chambers we inhabit than by direct experience.

The proliferation of misinformation and disinformation further complicates our understanding of the world. In the digital age, falsehoods can spread with alarming speed and reach, often masquerading as legitimate news and information. This constant exposure to fabricated narratives erodes trust in traditional sources of information and makes it increasingly difficult to discern fact from fiction.

Our relationship with technology also plays a crucial role in refracting reality. The devices we use to navigate the world, from smartphones to virtual reality headsets, mediate our experiences, shaping how we perceive and interact with our surroundings. While these technologies offer unprecedented opportunities for connection, information, and entertainment, they can also create a sense of detachment from the physical world, blurring the lines between the virtual and the real.

Consider the impact of augmented reality, which overlays digital information onto our physical surroundings. While this technology has the potential to enhance our understanding of the world, providing us with real-time data and insights, it also raises questions about how we distinguish between augmented reality and actual reality.

The rise of deepfakes, highly realistic AI-generated videos and audio recordings, further underscores the increasingly porous boundaries of reality. These sophisticated fabrications have the potential to deceive even the most discerning viewer, blurring the lines between truth and falsehood and eroding trust in visual and auditory evidence.

In this age of refracted reality, it becomes increasingly important to cultivate critical thinking skills, to question our assumptions, and to seek out diverse perspectives. Media literacy, the ability to critically analyze and evaluate information from various sources, emerges as an essential skill for navigating the complexities of the digital age.

Furthermore, fostering a sense of digital well-being becomes paramount. This involves being mindful of our technology consumption, setting healthy boundaries, and prioritizing real-world connections and experiences.

Ultimately, navigating the refracted realities of the 21st century requires a shift in mindset. It's about acknowledging the fluidity of perception, embracing critical thinking, and cultivating a healthy skepticism towards the information we encounter. By engaging actively and thoughtfully with the world around us, both physical and digital, we can begin to piece together a more nuanced and informed understanding of reality, even as it continues to refract and evolve before our very eyes.

A Kaleidoscope of Truth

Truth, once considered a singular, monolithic concept, now appears increasingly multifaceted, a shimmering kaleidoscope of perspectives, interpretations, and experiences. We inhabit a world where the pursuit of absolute truth often gives way to the acknowledgment of multiple truths, each valid within its own context, each contributing to the richness and complexity of human understanding.

The rise of postmodern thought in the late 20th century challenged traditional notions of objectivity and absolute truth, suggesting that meaning and understanding are fluid, contingent upon social, cultural, and historical contexts. This shift in perspective encourages us to move beyond simplistic binaries of right and wrong, true and false, and to embrace the inherent ambiguity and multiplicity of human experience.

The diversity of human cultures further enriches this kaleidoscope of truth. Each culture, with its unique history, values, and beliefs, offers a distinct lens through which to view the world. Practices and beliefs that might seem strange or irrational from one cultural perspective might hold deep meaning and significance within another. Recognizing and respecting these diverse perspectives is essential for fostering intercultural understanding and appreciating the multifaceted nature of truth.

Even within a single culture, individual experiences shape and color our understanding of truth. Our upbringing, our relationships, our triumphs, and our failures—all leave an indelible mark on our worldview, influencing how we interpret events, interact with others, and ultimately, define truth for ourselves.

The natural world, too, reveals the limitations of seeking a single, absolute truth. Scientific inquiry, while invaluable for understanding the physical universe, operates within the confines of its own methodologies and paradigms. What we consider scientific "truth" is constantly evolving, shaped by new discoveries, technological advancements, and evolving theoretical frameworks.

The arts, in their various forms, provide a powerful lens through which to explore the kaleidoscope of truth. Literature, painting, music, and film offer subjective interpretations of the human experience, inviting us to engage with multiple perspectives, challenge our assumptions, and embrace the emotional and psychological complexities of life.

This embrace of multiple truths does not, however, equate to the abandonment of all objective standards or the embrace of relativism. While acknowledging the subjective nature of perception and interpretation, we can still strive for honesty, integrity, and a commitment to seeking evidence-based understanding.

Cultivating critical thinking skills becomes paramount in navigating this complex landscape of truth. We must learn to evaluate information from diverse sources, to identify biases, and to engage in respectful dialogue with those who hold differing perspectives.

Ultimately, embracing the kaleidoscope of truth requires a shift in mindset. It's about moving beyond the pursuit of a singular, absolute truth and embracing the richness and complexity that emerge when we acknowledge the validity of multiple perspectives. By approaching the world with curiosity, empathy, and a willingness to challenge our own assumptions, we open ourselves to a more nuanced and enriching understanding of the human experience in all its multifaceted glory.

Embracing Multiplicity

The human experience, in its very essence, is a tapestry woven from a multitude of threads. We are complex beings, shaped by a confluence of factors:

biology and environment, culture and individuality, reason and emotion. To embrace this multiplicity, within ourselves and in the world around us, is to step into the fullness of human experience, to acknowledge the richness that arises from difference, and to cultivate a more compassionate and understanding approach to life.

Within each of us resides a multitude of selves. We are simultaneously children and adults, shaped by the echoes of our past and the aspirations for our future. We embody a spectrum of emotions, from joy to sorrow, anger to peace, often experiencing these seemingly contradictory feelings in rapid succession. To deny any aspect of this internal multiplicity is to deny a part of ourselves, to limit the fullness of our being.

Embracing our internal multiplicity requires a willingness to explore the hidden landscapes of our inner world. It involves acknowledging our shadows, those aspects of ourselves we may deem undesirable or unacceptable, and integrating them into a more holistic sense of self. It means honoring the full spectrum of our emotions, allowing ourselves to feel deeply and authentically, even when those emotions are challenging or uncomfortable.

This internal embrace paves the way for a more compassionate and understanding approach to the world around us. When we acknowledge the complexity and multiplicity within ourselves, we are more likely to recognize and honor the diversity of experiences in others. We become less likely to judge, to categorize, or to force others into neat, simplistic boxes.

The world we inhabit is a symphony of cultures, each with its unique history, traditions, and ways of seeing the world. Languages, with their subtle nuances and cultural embeddedness, shape our understanding of reality, influencing how we think, feel, and interact with our surroundings. Religious and spiritual traditions offer diverse paths towards meaning-making, providing frameworks for understanding our place in the universe and connecting with something larger than ourselves.

To embrace this external multiplicity is to cultivate an attitude of curiosity and openness. It means approaching different cultures with humility, seeking to understand the world through the eyes of those who have lived experiences vastly different from our own. It involves engaging in respectful dialogue, listening deeply to diverse perspectives, and recognizing that our own worldview is but one among many.

This embrace of multiplicity extends beyond the human realm. The natural world, in its breathtaking diversity, reminds us that life on Earth thrives on interconnectedness and interdependence. From the smallest microorganisms to the largest mammals, each species plays a vital role in maintaining the delicate balance of the ecosystem.

Embracing multiplicity, in all its forms, is an ongoing journey, not a destination. It requires constant self-reflection, a willingness to challenge our own biases, and a commitment to cultivating empathy and understanding. Yet, it is a journey well worth taking, for it leads us towards a richer, more compassionate, and ultimately, more meaningful way of being in the world.

Chapter 5: Shadows and Silhouettes

The Dance of Light and Dark

Life, in its most elemental form, is a dance of opposites. It is a rhythmic interplay of light and dark, joy and sorrow, creation and destruction. To embrace this dance is to embrace the totality of human experience, to recognize that even in the darkest of times, the seeds of light are present, and that within moments of profound joy, shadows inevitably linger.

The human psyche, much like the universe itself, is not a realm of absolutes but a dynamic interplay of contrasting forces. We are drawn towards the light, towards experiences of love, connection, and fulfillment. Yet, we also encounter darkness, those inevitable moments of loss, grief, and despair that cast long shadows across our lives.

To deny the darkness is to deny a fundamental aspect of our being. It is to resist the natural cycles of life and death, growth and decay, that shape the world around us and the landscapes of our souls. When we attempt to suppress or ignore the darker aspects of ourselves, they tend to fester beneath the surface, emerging in unhealthy and destructive ways.

Embracing the darkness, however, does not mean surrendering to despair or resigning ourselves to a life devoid of light. It means acknowledging the presence of suffering, both our own and that of the world, with courage and compassion. It means allowing ourselves to grieve losses, to feel the weight of sadness, and to sit with uncomfortable truths without judgment or resistance.

It is often within the depths of our despair that we discover an unexpected wellspring of strength and resilience. When we allow ourselves to fully experience the darkness, we often emerge from the experience transformed, our hearts cracked open in ways that allow for deeper compassion, greater empathy, and a more profound appreciation for the preciousness of life.

Moreover, it is within the interplay of light and dark that creativity often finds its most fertile ground. Artists, writers, musicians—throughout history, those who have dared to delve into the depths of human experience have often produced works of profound beauty and insight. The darkness, when embraced as a source of inspiration rather than something to be feared, can ignite our imaginations and fuel our creative fire.

This dance of light and dark is not limited to the internal landscape of our individual lives. It plays out on a grander scale in the world around us. Societies, much like individuals, experience periods of growth and prosperity, followed by inevitable periods of decline and upheaval. Natural disasters, conflicts, and economic downturns remind us of the fragility of life and the impermanence of all things.

Yet, even in the face of immense suffering and destruction, the human spirit reveals its remarkable capacity for resilience, compassion, and renewal. Acts of kindness, courage, and selflessness emerge from the ashes of tragedy, reminding us of the enduring power of the human spirit to choose hope even in the darkest of times.

To embrace the dance of light and dark is to embrace the totality of existence, with all its beauty and its pain, its joys and its sorrows. It is to recognize that life is not a linear journey towards enlightenment but a cyclical dance, a constant ebb and flow of opposing forces. And it is within this dance, this embrace of paradox and contradiction, that we discover the fullness of our humanity and the profound beauty of being alive.

Finding Meaning in Absence

Absence, in its various forms, is an inevitable part of the human experience. It weaves its way through our lives, leaving behind a complex tapestry of loss, longing, and the bittersweet ache of what once was. Yet, within these seemingly empty spaces, within the very heart of absence, lies a profound potential for growth, transformation, and the discovery of meaning.

We encounter absence in its most tangible form through the loss of loved ones. The death of someone dear to us creates a void, a physical absence that can feel impossible to bridge. Grief washes over us in waves, leaving us raw, vulnerable, and grappling with the profound silence left in their wake.

Yet, even in the depths of grief, the seeds of meaning can begin to sprout. We find solace in cherished memories, in the stories we share, and in the enduring impact our loved ones had on our lives. Their absence becomes a catalyst for reflection, prompting us to examine our own lives, our values, and the legacy we wish to leave behind.

Absence also manifests in the realm of relationships. The ending of a romantic partnership, the estrangement from a friend, or the distance that can grow between family members can leave us feeling unmoored, adrift in a sea of loneliness and longing.

These experiences, while undeniably painful, can serve as powerful catalysts for personal growth. They challenge us to examine our own needs, our patterns in relationships, and the ways in which we contribute to the dynamics that unfold between ourselves and others. The space created by absence provides an opportunity for introspection, for healing old wounds, and for cultivating a deeper understanding of ourselves and our role in the dance of connection.

The absence of things we yearn for—a desired career path, a creative pursuit that eludes our grasp, a longing for a different life than the one we find ourselves living—can also give rise to a profound sense of emptiness and longing. We may find ourselves grappling with feelings of inadequacy, questioning our choices, and wondering what might have been.

However, it is often within these spaces of unfulfilled longing that we discover hidden reservoirs of strength and resilience. We learn to navigate disappointment, to embrace uncertainty, and to find meaning and purpose even when our dreams take unexpected detours.

The absence of what we expect, of what we believe should be, can crack open our hearts and minds, creating space for new possibilities to emerge. It can lead us to unexpected paths, to hidden talents, and to a deeper appreciation for the gifts that are present in our lives.

Finding meaning in absence is not about denying the pain of loss or pretending that what is missing doesn't matter. It is about recognizing that even in the emptiness, even in the spaces where something or someone once stood, there is still beauty, still wisdom, still the potential for growth and transformation.

It is in the quiet moments of reflection, in the tender embrace of grief, in the space created by unfulfilled longings, that we often discover the most profound truths about ourselves and the world around us. For it is in the very heart of absence that we learn to cherish what we hold dear, to appreciate the ephemeral nature of life, and to find meaning in the ever-unfolding tapestry of human experience.

The Power of Negative Space

In the realm of art, negative space is the area surrounding the subject of an image. It's the "empty" space that allows the subject to breathe, to stand out, to be fully perceived. But this space, often overlooked, holds a power that extends far beyond the canvas, permeating our lives in profound and often unnoticed ways. Recognizing and harnessing the power of negative space, in our external environments and within the landscapes of our lives, can be transformative.

Consider a simple ink drawing. The artist, with deliberate strokes, outlines the petals of a flower, leaving the surrounding paper untouched. It's this untouched, "negative" space that defines the shape of the flower, allowing our eyes to recognize its delicate curves and graceful form. Without the negative space, the image would be a chaotic jumble of lines, the flower lost in the chaos.

This principle extends far beyond the artistic realm. In a bustling city, parks and open green spaces serve as negative space, providing respite from the concrete jungle. These pockets of "emptiness" are not mere voids; they are essential for our well-being, offering a place for reflection, connection with nature, and a restoration of our senses.

Within our homes, the power of negative space translates to mindful minimalism. A sparsely decorated room, free from clutter and excess, allows our eyes to rest and our minds to settle. It's not about starkness, but rather about creating breathing room, allowing the objects we choose to keep to hold more meaning and significance.

However, the power of negative space extends beyond the physical realm and into the very fabric of our lives. It's about recognizing the importance of "emptiness" – the pauses between our words, the silence between musical notes, the stillness we cultivate in meditation. These spaces, often perceived as passive or unproductive, are in fact fertile ground for reflection, creativity, and renewal.

In the relentless pace of modern life, we often equate busyness with productivity, filling our schedules to the brim and leaving little room for stillness. But it is in the negative space, in the moments of quiet contemplation, that we connect with our deepest selves, gain clarity of thought, and tap into our wellspring of creativity.

Consider the way ideas often strike us, not when our minds are racing from one task to the next, but in moments of quiet reflection—during a walk in nature, in the shower, or in the stillness of the early morning hours. These are the moments when we create the negative space for inspiration to take root and blossom.

Embracing the power of negative space also means recognizing the importance of saying "no" – to commitments that drain our energy, to relationships that no longer serve us, to the relentless pursuit of "more." Creating space in our lives often requires letting go, releasing what no longer serves us to make room for what truly matters.

This can be a challenging practice in a society that often glorifies busyness and productivity. But learning to discern the difference between what truly nourishes us and what simply clutters our lives is essential for creating a sense of balance, purpose, and well-being.

Chapter 6: Luminescence and Legacy

Leaving Our Mark

The human experience is a tapestry woven with countless threads, each life a unique strand contributing to the richness and complexity of the whole. As we navigate the terrain of our own existence, a fundamental longing resides within us: the desire to leave our mark on the world, to make a difference, to create something that will outlast our own fleeting time on this earth. This yearning to leave a legacy is not driven by ego or the pursuit of fame, but by a deep-seated desire to connect, to contribute, and to find meaning in the grand narrative of life.

The ways in which we choose to leave our mark are as varied as the individuals who make up the human family. For some, it may be through the creation of art, music, literature, or any form of creative expression that stirs the soul, ignites the imagination, and offers new ways of seeing the world. The artist pours their heart and soul onto the canvas, the writer bleeds ink onto the page, the musician channels emotions into melodies that resonate through time, touching countless lives they may never meet.

Others may find their purpose in acts of service and compassion, dedicating their lives to alleviating suffering, fighting for justice, or simply lending a helping hand to those in need. The doctor who heals the sick, the teacher who inspires young minds, the activist who challenges injustice—their legacies are etched in the lives they touch, the communities they uplift, and the ripple effects of their actions that extend far beyond their immediate sphere.

For some, leaving a mark may be as simple as creating a loving and supportive home for their family, nurturing the next generation, and passing down values, traditions, and stories that will shape the lives of their children and grandchildren. The impact of a parent's love, guidance, and unwavering support can reverberate through generations, creating a legacy of empathy, resilience, and enduring human connection.

It's important to remember that leaving our mark is not solely about grand gestures or monumental achievements. It's often the small, everyday acts of kindness, compassion, and integrity that leave the most lasting impact. A smile offered to a stranger, a listening ear lent to a friend in need, a moment of courage when it would be easier to stay silent—these seemingly insignificant actions can ripple outwards, creating a more compassionate and connected world.

The desire to leave our mark is not about seeking validation or recognition from others, but about living a life that is true to ourselves, aligned with our values, and in service to something larger than ourselves. It's about recognizing that we are part of something vast and interconnected, and that our actions, however small, have the power to make a difference.

In the grand scheme of things, our time on this earth is but a fleeting moment. Yet, the impact of our lives, the love we share, the kindness we offer, the contributions we make—these have the power to echo through time, shaping the world for generations to come. Leaving our mark is not about achieving immortality, but about embracing the preciousness of this life and striving to make it meaningful, not only for ourselves but for the world around us. For in the end, it is the love we give, the connections we forge, and the difference we make in the lives of others that truly define our legacy.

The Afterglow of Experience

Experiences, both grand and seemingly insignificant, wash over us like waves, shaping the shores of our lives. Some experiences crash upon us with great force, leaving their mark etched deeply in our memories. Others lap gently at our feet, their presence fleeting yet leaving behind a subtle shift in our being. It's tempting to chase after the next exhilarating wave, the next peak experience, but what about the moments that follow? The afterglow of experience, often overlooked in our relentless pursuit of novelty, holds a unique power – the power to deepen our understanding of ourselves, to solidify lessons learned, and to integrate the wisdom gleaned into the fabric of our lives.

Imagine standing at the summit of a mountain after a challenging climb. The air is thin, the world stretches out beneath you, and a sense of accomplishment washes over you. This is a peak experience, a moment of exhilaration and triumph. But what happens when you descend back down the mountain? The afterglow of that experience lingers – the memory of the struggle, the camaraderie of fellow climbers, the breathtaking views that expanded your sense of the world. It's in the afterglow that we process the experience, integrating the lessons of perseverance, resilience, and the sheer beauty of the natural world.

The afterglow is not merely about reminiscing about the past; it's about carrying the essence of the experience forward, allowing it to inform our present and shape our future. A conversation with a wise friend, a transformative travel experience, a moment of profound connection with nature – these experiences leave behind a residue of insight, inspiration, and a renewed perspective on our lives. It's in the quiet moments of reflection that follow, as we revisit these experiences in our minds, that their true power is revealed.

One way to cultivate the afterglow of experience is through the practice of journaling. Taking the time to write down our thoughts and feelings about an experience, whether it's a joyful celebration or a challenging life event, allows us to process our emotions, gain clarity, and uncover hidden layers of meaning. Through the act of writing, we create a tangible record of our experiences, a tapestry of words that we can revisit and reflect upon over time.

Another powerful way to extend the afterglow is through sharing our experiences with others. When we share our stories, our vulnerabilities, and our triumphs, we create a space for connection, empathy, and shared understanding. The act of sharing not only deepens our own connection to the experience but also allows others to glean insights, inspiration, and perhaps even solace from our journeys.

However, the afterglow is not always about positive experiences. Difficult times, moments of loss, grief, or failure, also leave their mark on us. The afterglow of these experiences can be painful, but it can also be profoundly transformative. It's in the processing of these challenging emotions, in the quiet contemplation of what we've lost or what could have been, that we often discover our greatest reserves of strength, resilience, and compassion.

The afterglow of experience is a reminder that life is not simply a series of events, but a continuous journey of growth, transformation, and integration. It's about savoring the peak moments, yes, but also about honoring the spaces between, the quiet reflections that allow us to make sense of it all. For it's in the afterglow, in the subtle shifts in our perceptions and the quiet integration of wisdom gained, that we truly weave the tapestry of our lives.

Passing the Torch

The cycle of life, in its relentless yet beautiful rhythm, carries with it an inherent truth: we are all part of a continuum, a chain of interconnected lives stretching back generations and reaching far into the future. Each of us holds a unique place in this chain, carrying forward the wisdom, traditions, and values passed down to us, while simultaneously shaping the legacy we will one day pass on. This act of passing the torch, of ensuring that the flame of knowledge and experience continues to burn brightly for generations to come, is a profound responsibility and an essential aspect of the human experience.

Throughout history, the passing of the torch has taken many forms. In ancient cultures, elders held a revered position, their wisdom and life experiences sought after and cherished by younger generations. Stories were shared around campfires, skills were passed down through apprenticeships, and the traditions that held communities together were carefully preserved and transmitted. This intergenerational exchange ensured that the accumulated knowledge and wisdom of the past were not lost, but rather served as a foundation for the future.

In the realm of art, music, literature, and countless other creative pursuits, the passing of the torch is evident in the lineage of masters and apprentices. Think of the renowned artists whose studios were once bustling with eager students, each one absorbing the techniques, perspectives, and philosophies of their mentors. Think of the musical traditions passed down through families, each generation adding their unique voice to the symphony of their heritage. This passing of the torch ensures that creative flames continue to ignite, illuminating the world with beauty, insight, and the power of human expression.

However, the act of passing the torch is not limited to formal settings or specific disciplines. It happens in the everyday moments of our lives – in the kitchen, where grandparents teach grandchildren age-old recipes, their hands moving in a familiar dance of tradition and love. It happens on front porches, where stories of hardship overcome and dreams realized are shared, weaving a tapestry of family history that connects generations. It happens in the simple act of listening, truly listening, to the experiences and perspectives of those who have walked the path before us, honoring their wisdom and carrying their stories in our hearts.

The responsibility of passing the torch is not solely about preserving the past; it's about empowering the future. It's about recognizing that each generation has a unique contribution to make, a fresh perspective to offer, and a responsibility to build upon the foundations laid by those who came before. It's about encouraging young minds to question, to explore, to innovate, and to ultimately find their own voices and make their own indelible marks on the world.

In a world often obsessed with novelty and the pursuit of individual achievement, the act of passing the torch reminds us that we are part of something much larger than ourselves. It reminds us that our lives have meaning and purpose beyond our own fleeting time on this earth. It reminds us that the greatest gifts we can offer future generations are the gifts of knowledge, experience, and the enduring flame of human connection.

As we navigate the complexities of our own lives, let us remember the importance of both receiving and passing the torch. Let us seek out the wisdom of those who have come before us, absorbing their stories, honoring their legacies, and carrying their spirits within us. And let us create opportunities to share our own experiences, insights, and hard-won wisdom with the generations that follow, ensuring that the flame of knowledge, compassion, and the indomitable human spirit continues to burn brightly, illuminating the path for generations to come.

Chapter 7: Conversations in the Light

Dialogue and Discovery

Dialogue, in its purest form, is a dance. It's a delicate interplay of thoughts, emotions, and perspectives, weaving together a tapestry of understanding. We engage in dialogue not just to exchange information, but to connect, to learn, and to grow. It's through these conversations, often spontaneous and unscripted, that we stumble upon unexpected insights and uncover hidden truths.

Think about those moments when a simple conversation sparked a brilliant idea. Perhaps it was a casual chat with a colleague that led to a breakthrough solution at work, or a heart-to-heart with a loved one that shifted your entire perspective. These moments of discovery are rarely planned; they emerge organically from the fertile ground of genuine dialogue.

But how do we cultivate this kind of enriching conversation? How do we move beyond the superficial and delve into the realm of meaningful exchange? It starts with a willingness to listen – truly listen – with an open mind and an empathetic heart. When we approach a conversation with the intent to understand rather than to simply reply, we open ourselves up to a world of discovery.

Imagine yourself sitting across from someone with a completely different viewpoint than your own. Instead of preparing your rebuttal while they speak, try to truly understand their perspective. Ask clarifying questions. Seek out the "why" behind their beliefs. You might not agree with everything they say, but by approaching the conversation with curiosity and respect, you open the door to empathy and understanding.

This willingness to listen is only one side of the dialogue coin. The other side, equally important, is the courage to speak your truth. Authentic dialogue requires vulnerability. It requires us to share our own thoughts and feelings, even when they differ from the norm. This doesn't mean bulldozing others with our opinions, but rather expressing ourselves with clarity and respect, inviting others to see the world through our lens.

This dance between listening and speaking, between receptivity and expression, is what fuels the fire of discovery. It's in the space between our words, in the silences that invite reflection, that we often find the most profound insights.

Consider the role of silence in your own conversations. Do you allow for pauses and moments of contemplation? Or do you rush to fill every gap with words? Silence can be uncomfortable, but it's often in those moments of quiet reflection that our thoughts crystallize and new ideas begin to form.

Dialogue, in its truest sense, is a journey of co-creation. It's a shared exploration where each participant brings their unique experiences and perspectives to the table. It's not about winning or being right; it's about building bridges of understanding and uncovering shared truths.

So, the next time you find yourself engaged in conversation, remember the power of dialogue. Approach the interaction with an open mind, a willing heart, and a thirst for discovery. Listen deeply, speak your truth, and embrace the silences in between. You might be surprised by what you uncover – about yourself, about others, and about the world around you.

Illuminating Our Shared Humanity

We are all storytellers. From the moment we learn to speak, we weave narratives that shape our understanding of the world and our place within it. Our stories, though unique in their details, are often woven from common threads – love, loss, joy, fear, hope. It is in the sharing of these stories, in the recognition of our shared human experience, that we discover the profound truth of our interconnectedness.

Think about a time you felt truly seen and understood by another person. Perhaps it was a moment of shared vulnerability, where you dared to peel back the layers of your carefully constructed persona and reveal the raw, messy truth beneath. Or maybe it was a moment of shared joy, where laughter flowed freely and the weight of the world seemed to momentarily lift. In these moments of genuine connection, we catch glimpses of our shared humanity, recognizing that beneath the surface of our differences lies a common ground of experience.

One powerful way we illuminate this shared humanity is through the act of empathy. Empathy is not simply feeling sorry for someone; it's the ability to step outside of our own experience and into the shoes of another. It's about seeking to understand the world through their eyes, to feel the world through their heart. It's a conscious choice to bridge the gap between "me" and "you," to recognize that their story, however different from our own, is no less valid, no less real.

Cultivating empathy requires us to challenge our own biases and assumptions. It asks us to confront our own prejudices and to acknowledge the ways in which our own experiences have shaped our worldview. This can be uncomfortable work, but it is essential work if we are to truly see and connect with those who are different from us.

Storytelling, in its many forms, can be a powerful tool for fostering empathy. When we engage with stories, whether through books, films, music, or personal narratives, we open ourselves up to the experiences of others. We step into their world, feel their joys and sorrows, and emerge with a deeper understanding of the human condition.

Consider the impact of a story that has stayed with you long after you finished reading or watching it. Perhaps it was a fictional account that transported you to a different time and place, allowing you to experience life through the eyes of someone vastly different from yourself. Or maybe it was a personal story shared by a friend or family member, a story that gave you a glimpse into their struggles and triumphs, their hopes and fears. These stories have the power to stay with us, to shape our perspectives, and to expand our capacity for empathy.

But illuminating our shared humanity goes beyond simply understanding and empathizing with others. It also requires us to recognize and challenge the systems and structures that perpetuate division and inequality. We must acknowledge the ways in which our society often prioritizes certain voices and experiences over others, and actively work to create a more just and equitable world for all.

This means speaking up against injustice, even when it's uncomfortable. It means using our voices to amplify the stories of those who have been marginalized and silenced. It means being willing to listen to and learn from those who have different experiences than our own.

Illuminating our shared humanity is a continuous journey, not a destination. It's a process of learning, unlearning, and relearning. It's about recognizing our interconnectedness, celebrating our diversity, and working together to create a world where everyone feels seen, heard, and valued.